DIY OFF-GRID SOLAR POWER FOR BEGINNERS
(2020 EDITION)

A step-by-step instruction to Power Mobile Homes, Camper's Vans, RVS and Boats from the sun

LARRY BARONE

Copyright

Larry Barone
ISBN: 9798665111162
ChurchGate Publishing House
USA | UK | Canada
© Churchgate Publishing House 2020

All rights reserved. No part of this publication may be reproduced, stored in a retrieval system or transmitted in any form or by any means, electronic, mechanical, photocopying, recording, and scanning without perm-ission in writing by the author.

While the advice and information in this book are believed to be true and accurate at the date of publication, neither the authors nor the editors nor the publisher can accept any legal responsibility for any errors or omissions that may be made. The publisher makes no warranty, express or implied, with respect to the material contained herein.

Printed on acid-free paper.

Printed in the United States of America
© 2020 by Larry Barone

Contents

Copyright i
Introduction 1
Chapter One 3
A Sneak Peak on Electricity 3
Origin of electricity 3
History of electricity 4
Types of electricity 5
What is electricity used for? 5
How is electricity manifested? 6
How is electricity generated? 7
How is electricity transmitted? 9
Electric conductivity 9
What Is Renewable Energy 10
Exploring the skeletal system of solar power 16
What you need to know about mobile solar power 18
Initial Cost 19
Components of Mobile Solar Power 19
Types of solar panels for mobile solar power systems 21
Charge Controller 23
Batteries 24
Inverters 26

A Brief Guide to Installing Mobile Solar power 27

What is the best Panel For You? 28

Why you should use mobile solar power 30

Benefits of using mobile solar power systems for vans 30

CHAPTER TWO 32

Introduction to mobile solar power 32

Advantages of using mobile solar power 35

Explaining The Components of Mobile solar power system 36

Purchasing Your Mobile Solar panel 41

Types of solar cells, monocrystalline vs polycrystalline 46

Requirements for installing a mobile solar power system 47

Chapter Three 49

Selecting a solar panel 49

Sizing You Solar Panel and Battery 49

Calculation of the daily consumption of an electrical installation of a van 49

Summer Consumption 50

Winter Consumption 51

Sizing of the solar panel in a van 51

How to size the capacity that your battery must have 52

How to Install a Mobile Solar Power 54

Tools to use for the installation of a solar panel in a van 54

Function test of the solar panel plate 55

How to connect solar panels in series or parallel 56

Connect solar panels in parallel 58

Connect panels in series and parallel (mixed) 59

Is it better to connect solar panels in series or parallel? 59

What differences exist between the types of connectors (MC4 and SAE) 60

Best practice for a water-tight cable installation 61

Chapter Five 63

Introduction into boats mobile power system 63

Advantages of using mobile solar power for boats 64

How to install mobile solar power for boats 67

Chapter Six 70

Introduction **to Small Homes Mobile Power System** 70

Advantages of Using Mobile solar Power System for Tiny Homes 70

Components of Tiny Homes Mobile Solar Power System 72

How to Install Mobile Solar Power for Tiny Homes 77

Positioning the Solar Panels 77

Chapter Six 82

A simple guide on how to maintain a mobile solar power system 82

How is the maintenance of the solar panels carried out? 83

Troubleshooting Your Mobile Solar Power System 85

Types of problems that can occur on a solar panel 86

Conclusion 89

About the Author 91

Introduction

A mobile solar power system is the perfect ally for the energy efficiency of your camper van, small home or boat. Photovoltaic solar energy as it is often called is a clean energy source when it comes to impact on the environment. It produces electricity of renewable origin from solar radiation using a semiconductor device, the photovoltaic cell.

Mobile solar installation in a well-dimensioned camper allows you to have more freedom on your route since your van becomes self-sufficient. For this reason and others that we are going to explain, solar panels are a key part of the electrical installation of a van. This is even important if you are going to lead a nomadic life. Parked or driving, cloud or sun, it does not matter, as long as there is daylight, solar panels continue to produce electricity.

With this guide, we want to help guide you on how to install an off-grid mobile solar power system on your own. This guide is also going to explain how to size the solar panel and the battery you need for the solar installation of your van.

Chapter One

A Sneak Peak on Electricity

Electricity is a set of physical phenomena that arise from the existence and flow of electrical charges. In other words, they depend on one of the fundamental properties of matter. This property is a consequence of the relationships between electrons (subatomic particles that have electromagnetic charge, conventionally designated as negative (-)).

What we commonly refer to as electricity is not exactly electricity. The latter is a consequence or application of the flow of electrons through a conductive material (electric current). It is also a highly versatile form of energy, essential in contemporary human life.

Electricity is part of electromagnetism, one of the four fundamental natural forces (along with gravity, weak and strong nuclear forces). Electricity today is generated, transported, stored, and consumed. This means it is transformed into other types of energy used by man: thermal, kinetic, chemical, light, etc.

Origin of electricity

Electricity is a natural interaction that concerns the atoms of which matter is made. When an atom loses or gains electrons, it changes its electrical state to obtain a charge (positive if it loses electrons, negative if it gains them).

Different phenomena can induce this effect in the matter, generating electrical imbalances (electrification) that can later generate electrical currents. For example, rubbing some fabrics (such as wool) can generate noticeable static electricity.

History of electricity

Thanks to William Gilbert with whom the discovery of the fields of electricity began. Since ancient times, humanity has sensed the presence of electricity, observing it in nature. However, its formal study began with the Scientific Revolution of the 17th and 18th centuries, and only in the 19th century could it be applied for domestic and industrial uses.

The earliest handling of electricity in the 17th century was little more than parlor attractions, with which the bourgeoisie was impressed. But at that time the English philosopher William Gilbert dedicated himself to the study and differentiation of the phenomena obtained by rubbing amber (static electricity) and that of magnetite (magnetism) despite being similar in their attraction to small objects. This was the beginning of the discovery of the fields of electricity and magnetism, the relationship of which would be understood much later.

The great scientists in charge of understanding electricity were mostly from the 18th century: Cavendish, Du Fay, van Musschenbroek, and Watson, as well as Galvani, Volta, Coulomb, and Franklin. Already in the early 19th century Ampère, Faraday, and Ohm joined, as did James Clerk Maxwell, who first formulated the unifying equations of electricity and magnetism.

Types of electricity

There are two fundamental types of electricity:

Static electricity: It is generated around a load at rest or stillness, that is, it does not move or flow. For example, when a piece of amber is rubbed with a ball of wool or dry cloth, an electronic imbalance occurs in the amber, giving it an electrical charge. This charge resides in the amber until it is balanced in some way.

Dynamic electricity: It is the electricity generated around a moving charge, that is, the flow of an electric charge: electric current. This requires a permanent source of electricity that makes electrons flow through the body of a conductive material, and can be made to flow to other atoms. This makes it truly useful.

What is electricity used for?

Electricity is a very powerful and very versatile force, which serves to power all kinds of devices and chemical reactions.

So it can be converted into other forms of usable energy. For example, it can be used to generate heat through resistors, allowing you to heat a room or even cook food. It is also used to generate light through bulbs or to run an engine and generate motion. Electricity powers electronic devices capable of endless purposes, from ringing a bell to carrying out arithmetic operations.

How is electricity manifested?

Electric charges generate electromagnetic fields around them. Electricity manifests itself in a set of phenomena and physical properties:

- **Electric charge.** Some substances have or can have an electric charge, depending on the behavior of their electrons. This determines their electromagnetic interaction.

- **Electric current.** The flow or displacement of electrons through a conductive material is called electric current.

- **Electric field.** Electric charges generate electromagnetic fields around you, affecting any other charges in your vicinity.

- **Electric potential.** The ability of an electric field to perform work.

- **Magnetism.** Electricity and magnetism are closely related. So much so that electric current generates magnetic fields, and magnetic fields that vary over time produce an electric current.

How is electricity generated?

Electricity is generated in power plants of various kinds, usually through electromagnetic generators that through movement sustain a difference in electrical potential between two points.

This movement is generally driven by rising water vapour or other hydraulic forces: wind, falling water, etc., acting on a turbine. Thus, generators move a magnetic field (such as a magnet or electromagnet) on conductive materials. Below are a few energy sources for electricity

Thermal energy: Thermal energy is called the deliberate force in the form of heat, it can be obtained from nature or the sun through an exothermic reaction, such as the combustion of some fuels by a nuclear fusion reaction using electric energy by reasoning or by reasoning, as a residue of other mechanical or chemical processes.

Wind energy: It is the energy obtained from the wind, that is, kinetic energy generated by the effect of air currents and which is used in other ways useful for human activities.

Geometric energy: It is the energy that can be obtained by taking advantage of the heat from the interior of the earth.

Nuclear energy or atomic energy: It is the energy that is released spontaneously or artificially in nuclear reactions.

Static energy: It is quiet energy, electrons at rest; for it to occur there must be a flexion between two bodies, this energy is one of the most dangerous when it comes to electronic components. This energy can easily burn electrical appliances.

Hydroelectric energy: Harnessing the potential energy accumulated in the water to generate electricity is a classic way of obtaining energy. About 20% of the electricity used at the source. It is therefore renewable energy but not an alternative one.

Solar energy: It is the energy obtained through the training of light and the heat obtained and emitted by the sun, the solar radiation reaching the earth can be used through the heat it produces through the absorption of radiation.

Hydraulic energy: Hydraulic or hydraulic energy is the one that is obtained from taking advantage of the kinetic and potential energy of the water current, waterfalls, or tides. It is a type of green energy. When its environmental impact is minimal and it uses the hydraulic force without re-growing it, otherwise it is considered only a form of renewable energy.

How is electricity transmitted?

Another vital matter in the management of electricity is its transmission from the generation source to the place of consumption. For this, conductive material wiring is available. But there is a dilemma, the greater the distance, the greater the loss of electrical charge. This is because even conductive materials are resistant to some small extent.

High-voltage lines are used to solve this dilemma since by tensioning, the current it is possible to cover more distance with fewer losses due to heating and electromagnetic effects. However, high voltage has two problems. On one hand, it is useless in domestic terms, and on the other hand, it is risky. The invention of the transformer was key to solving these problems: high voltage is used for transport and low voltage for consumption at the destination.

Electric conductivity

Electrical conductivity is called the ability of matter to allow the passage of electrical charges. It is a magnitude contrary to resistivity.

Depending on their nature, the materials may be:

Conductors. Those that allow the transit of electricity to all points on its surface, once exposed to it. The best-known conductors are metals and some versions of carbon, as well as most salts. In this process, part of the electrical charge is usually lost and heat is generated.

Dielectric or insulating. Materials that do not allow the passage of electricity, such as glass, bakelite, or plastics. For this reason, they are used as protectors and cable covers.

Semiconductors. Those that allow the passage of electricity in certain conditions of temperature, pressure, etc., while in others they act as an insulator. Examples of this are silicon, cadmium, or germanium.

Superconductors. In this case, the materials allow the passage of electricity without incurring any wear or loss of load, as long as they are in certain temperature conditions. This is what happens with tin and aluminium when they cool below their critical temperature.

What Is Renewable Energy

Renewable Energy is the energy obtained from theoretically inexhaustible natural sources. This is often due to the immense amount of energy they contain or their ability to regenerate themselves by natural means. Renewable energies are considered wind, geothermal, hydroelectric, tidal, solar, wave, biomass, and biofuels. These renewable energy sources can be non-polluting or clean, and other pollutants.

The non-polluting renewable energy sources are:

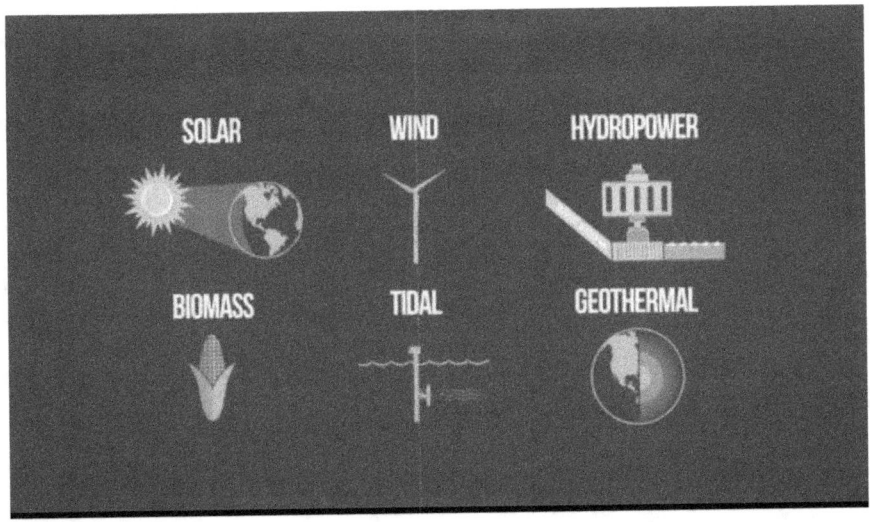

Wind energy

Wind energy is the energy obtained from the wind, using the kinetic energy generated by air currents. Wind turbines convert the kinetic energy of the wind into electrical energy using propellers that rotate a central axis connected to an electric generator.

Photovoltaic Solar Energy

The energy emitted by the sun can be mainly converted into two forms of energy: thermal (heat) and electrical. Thermal solar energy consists of collecting energy from the sun through solar panels and converting it into heat, to obtain hot water. The most developed source of solar energy is photovoltaic solar energy, which uses photovoltaic panels to collect the energy that is then converted into electrical energy.

Geothermal energy

Geothermal energy is obtained by taking advantage of the heat from the Earth's interior. In some areas of the planet, part of the Earth's internal heat reaches the surface of the Earth's crust. This energy can often be used to drive electric turbines or to heat.

Hydraulic or hydroelectric power

Hydraulic or hydroelectric power is generated by rivers and streams of freshwater. The accumulated potential energy in waterfalls can be transformed into electrical energy. Hydroelectric plants use the energy of the rivers to run turbines that run an electric generator. It is a clean and indigenous energy source, depending on weather conditions, and large infrastructure investments.

Marine energy

The energy of the seas and oceans refers to the energy produced by waves, tides, salinity, and ocean temperature differences.

The movement of water in the oceans creates kinetic energy or energy in motion, which can be used to generate electricity.

The main types are:

- Tidal waves, generated by the seas and oceans.
- Tidal waves, generated by taking advantage of the temperature difference between the sea surface and the depths.
- Osmotic power, generated by the difference in salt concentration between seawater and river water.

Biomass

It consists of the use of organic matter originating from a biological process as an energy source. Plants transform energy from the Sun into chemical energy through photosynthesis, and part of that chemical energy is stored in the form of organic matter. The chemical energy from biomass can be recovered and used as fuel by incineration (wood, solid plant matter, or urban waste), or converted into bioethanol, biogas, or biodiesel.

It is considered renewable energy because as long as the vegetables that produce them can be grown, they will not be depleted. It is also considered cleaner than its fossil equivalents because in theory the carbon dioxide emitted from combustion has previously been absorbed by transforming itself into organic matter through photosynthesis. In reality, the amount of carbon dioxide previously absorbed and the amount emitted in the combustion is not equivalent, since, during the sowing, harvesting, treatment and transformation processes, energy is also consumed, with its corresponding emissions.

Energy revolution and environmental impact of Renewable Energy

Renewable energy is at the centre of the current energy revolution. All of man's activities produce some degree of environmental impact, including the generation of energy from his various sources. Geothermal energy can be very harmful if heavy metals and greenhouse gases are dragged to the surface. Wind produces a negative visual impact on the landscape, low-frequency noise, and can be a bird trap.

The least aggressive hydraulics is the mini hydraulics since the large dams cause loss of biodiversity, generate methane from non-withdrawn plant material, cause pandemics such as yellow fever, or dengue fever. Large dams also flood areas with cultural heritage or landscape. Sometimes, it generates the movement of entire populations and increases the salinity of the wholesale riverbed jerseys.

Solar energy is among the least aggressive due to the possibility of its distributed generation, except for photovoltaic and thermoelectric electricity produced in large grid-connection plants, which generally use a large area of land. The tidal wave has been discontinued due to the high initial costs and the environmental impact that they entail. The energy of the waves together with the energy of the marine currents usually has a low environmental impact since they are usually located on rough coasts.

Biomass energy produces pollution during combustion by CO_2 emission but is reabsorbed by the growth of cultivated plants. It needs arable land for its development, decreasing the amount of arable land available for human consumption and livestock. Also, it comes with the danger of increasing the cost of food and increasing the production of monocultures.

It is true that biomass actively stores carbon from carbon dioxide, forming its mass with it and growing while releasing oxygen again. When burned, it combines carbon with oxygen again, forming carbon dioxide again. In theory, the closed cycle would yield a zero net balance of carbon dioxide emissions, as the emissions resulting from combustion remain fixed in the new biomass. In practice, polluting energy is used in planting, harvesting, and transformation, the balance being negative.

All biofuels produce more carbon dioxide per unit of energy produced than fossil equivalents. Geothermal energy is not only very geographically restricted, but some of its sources are considered pollutants. This is a consequence of which the extraction of groundwater at high temperatures generates the drag to the surface of unwanted and toxic salts and minerals.

Exploring the skeletal system of solar power

Solar energy, in its most basic form, is the radiation generated by the sun that reaches Earth in the form of light and heat. This energy can be converted to a form usable by artificial means.

Types

The two main types of solar energy are solar electricity and thermal solar energy. For the first, the energy from the sun is converted into electricity, while the second is used to heat a fluid at a high temperature. The wind is caused by variable warming of the Earth's surface by the sun, therefore technically wind energy is the third type of indirect solar energy.

Residential conversion

Solar energy is normally converted for a residential level with photovoltaic cells to produce electricity, and with non-concentrating thermal collectors for heating. According to the Department of Energy (DOE), residential solar power systems can be purchased to connect to the local power grid.

Large-scale conversion

Large-scale conversion of solar energy for commercial and industrial purposes typically employs parabolic troughs and solar dishes, which concentrate it to generate heat to drive turbines. For large-scale conversion of solar thermal energy, thermal concentrator collectors are used that can generate extreme heat.

Pros and cons

Solar energy produces no pollution for power generation and can help limit residential and commercial energy costs. However, the DOE believes that some large-scale maintenance systems are inappropriate and could cause negative environmental impacts.

Environmental limitations

Solar power can be obtained in most places on a limited scale, but for large-scale production, a good amount of constant sunlight is required, which is usually only found in desert regions. The Department of Energy reported that most of the large solar power plants are located in California.

WHAT YOU NEED TO KNOW ABOUT MOBILE SOLAR POWER

First of all, we must assess whether it pays to install mobile power systems or not. If you are one of those who do a few km a day and end up sleeping in a campsite where you can connect to the electricity grid, you surely don't need this type of installation. If, on the other hand, you spend several days in the same location and do not usually stay overnight in campsites, the things you need to know are:

Know the space you have since the greatest limitation of these facilities is the usable space on the ceiling of your van.

Know the power we need based on our consumption.

Types of trips and duration of the trip.

The maintenance of the panels is not very complicated, in the beginning, you should check that they all work correctly; Perhaps, sometimes, some panels can break due to hail, for example, or get dirty by bird droppings, or dust. Remember that a dirty solar panel loses its performance by approximately 40%.

Initial Cost

The initial cost of setting up a mobile solar power system depends primarily on your power needs. To obtain the cost of the system, you have to first determine the number of solar panels you will require for your van. This can be done by first calculating your total power consumption in watts. Solar panels come in various watt sizes ranging from 100 to 360 watts solar panels. Setting up a mobile solar power system may require as much as $1,000 depending on your electrical needs.

Components of Mobile Solar Power

As earlier stated, a photovoltaic solar power system is the best option if we want the electrical installation of our van to be self-sufficient. Regardless of the type of panel we choose in our installation, to understand this, we must know each component and its functionality.

Currently, there are different photovoltaic solar power systems. We will focus on those who do not have direct contact with the electricity grid, that is, isolated solar panel systems. Let's go on to list the different equipment of an isolated photovoltaic system.

Photovoltaic solar panel

The solar module or solar panel is the essential component in a mobile solar installation. Because it is in charge of capturing and converting solar energy into direct electric current. Installation can be composed of several solar panels to achieve the consumption power that is needed in the installation of our van.

Solar panels are made of silicon cells connected forming a network. These cells are covered by glass that protects them from inclement weather such as hail, rain, humidity, etc. As we have said, solar panels need to be prepared to withstand weather wear but they must also be prepared to withstand strong winds. The supports are a fundamental element for solar installations both for the orientation of the panel and for its fixing. More than anything so that it does not get reversed when we go on the road and can cause damage or accidents.

Types of solar panels for mobile solar power systems

There are several types of solar panels depending on the material used in your photovoltaic cells. This makes its efficiency and cost different and becomes a fundamental factor in choosing the plate that best suits our needs. The only stumbling block here is undoubtedly the initial economic investment, an investment that will later be fully profitable.

Monocrystalline solar panel

Monocrystalline panels work best in cold or cloudy climates. These solar panels absorb radiation much better but withstand overheating less and their performance is lower when this happens. These photovoltaic solar panels are the most efficient since they convert around 25% of sunlight into electricity. Also, monocrystalline panels have a longer useful life compared to other solar panels. Normally manufacturers' warranties are 2 years, although being made of crystalline silicon, a high-quality material, they usually last well over 25 years. On the negative side, monocrystalline solar panels are the most expensive, but they are always a great investment.

Polycrystalline solar panel

Polycrystalline panels are designed for hot climates as they absorb heat faster and perform better in overheated conditions. In other words, its photovoltaic cells produce more energy in conditions of high temperatures. Also, its production involves a smaller amount of silicon waste compared to monocrystalline plates. The manufacture of polycrystalline solar panels is more economical and this can be seen in their lower prices compared to monocrystalline ones.

Flexible solar panel

As we said before, the operation of solar panels has changed very little over the years. What has happened is that technology has allowed us to evolve and improve some aspects. Flexible solar panels are a clear result of this evolution.

These plates reduce to the maximum the necessary material to obtain electrical energy. This reduction also allows silicon cells to capture solar radiation in the same way. Flexible solar panels can be better adapted to spaces with more special shapes without the need for supports, such as the curved roof of a van.

As a drawback, we find that these flexible plates are more expensive if we compare them with other solar panels. As its production is focused on boats, campervans, and boats, its production is lower and this is evident in the price.

Charge Controller

The charge controllers are responsible for managing and dose energy to the batteries. They are a very important component in a photovoltaic system since, in addition to lengthening the useful life of the batteries; they avoid overcharges and over-discharges of the installation. The charge regulator in a solar installation is the union between the solar panels with the other elements of the installation.

In addition to setting the value of the voltage at which the solar installation of our van works, it allows the output of the continuous voltage with which the batteries are charged. An important thing to keep in mind is the polarity of the regulators when connecting them. If we reverse the polarity of this equipment, we will melt the solar installation.

Batteries

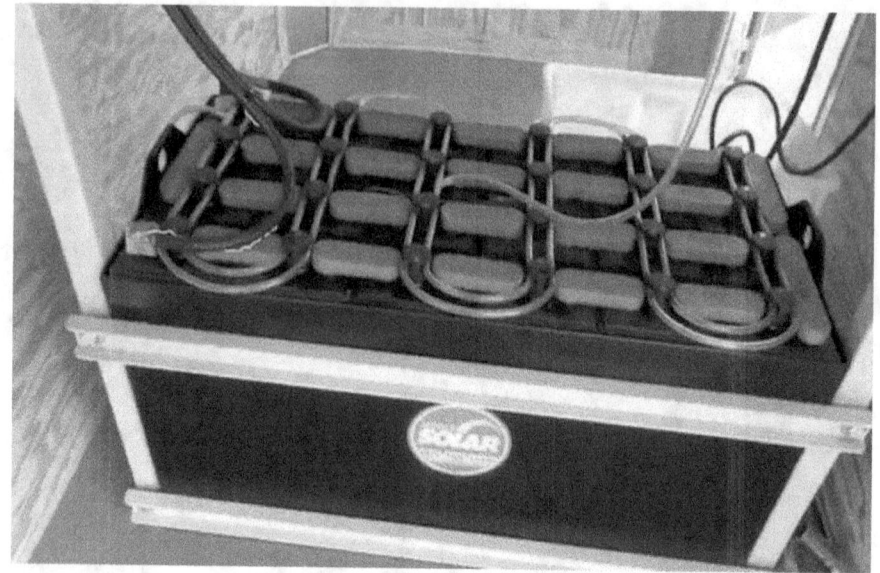

Batteries in photovoltaic solar installations are the element that allows installations to be autonomous. Accumulators of electrical energy make it possible to use that energy at another time. In the case of mobile solar systems, the interesting thing is that, in addition to the fact that the second battery has a large capacity for storing energy, it can suffer repeated discharges. The function of the batteries can be summarized as:

- Store energy for a few days.
- Provide a power trip at a certain time.
- Set the working voltage of the installation.

In the market, when we talk about automotive, we stick to three types of batteries: gel, acid, and AGM.

AGM batteries

The AGM batteries are very useful for solar installations in vans as they support a large number of charge and discharge cycles. These lead-acid batteries have a long service life (5 to 8 years), are completely sealed, so they do not require maintenance or ventilation and do not emit gases.

160Ah AGM battery

AGM batteries or accumulators can easily withstand the starting peaks of small motors and do so by providing high currents. Also, as they are compact in manufacturing, they withstand vibrations very well. Without a doubt, AGM batteries are an excellent option to install them as a secondary battery in the installation of your van. We use a secondary AGM battery in our installation.

Acid batteries

Acid batteries are the most common, the cheapest, and the ones that we can find in automobile engines used for starting. They are composed of a mixture of water and sulphuric acid and unlike AGMs, they cannot tip because this mixture is highly corrosive if it is spilled.

They are not recommended to be used as secondary batteries because during the discharge cycles they emit fumes that are harmful to health and cannot withstand full discharge cycles.

Gel batteries

They have an operation similar to that of AGM batteries, are of higher quality than these, and have a useful life that goes beyond 10 years. Gel batteries have a more stable voltage during discharge, ideal for use with inverters.

They differ because, instead of containing liquid sulphuric acid like other batteries, they contain it in the form of a gel. This is an advantage when it comes to handling since if they lean or lie down they will not spill either. Like AGMs, they are completely waterproof, maintenance-free and do not emit harmful gases.

Inverters

The function of the inverters is to convert direct current to alternating current. Some give continuous voltage, but if you choose this type of inverter you will not be able to use household appliances or electronic devices such as those at home.

For this, we need a pure wave inverter and in this way obtain a sine wave that is the one used by electronic devices. This voltage must be the same as that used by the electrical network, which is 220-230V of effective value and a frequency of 50 Hz. Basically what they provide us with is the 230v alternating current that we have in our homes.

Inverters are a very useful element in solar installations in vans that can be connected (or not) to the electrical grid.

A Brief Guide to Installing Mobile Solar power

In some shops, you can buy all the necessary products for the installation separately. However, there are configured solar kits for vans based on your needs. Buying the set will be cheaper.

To install your mobile solar power started to follow these basic instructions;

Glue the solar panel with silicone to the ceiling and pass the cables inside

Connect the regulator to the battery and then the plates to the regulator

Use a fuse between the regulator/inverter and the battery and another advisable between the regulator/inverter and the battery to avoid overcharges

Use the proper wiring. Please note the length and thickness.

What is the best Panel For You?

The first thing to know is the types of solar panels that exist and can be used in a camper van. They can be rigid or flexible, and monocrystalline or polycrystalline. Let's look at each type.

Flexible solar panel

They are extra thin solar panels, with a thickness of less than one centimetre making them the ideal solar panels for the roof of camper vans. They are also very light solar panels, so it does not add extra weight to the van.

However, they are very fragile plates, since they do not have a glass layer for protection, in addition to the fact that photovoltaic cells are not flexible on their own and can break. As the lack of ventilation (are bonded to the sheet), high-temperature performance is severely affected. The price of these solar panels is much higher than rigid solar panels.

Rigid solar panel

They are the "typical" solar panels that we see everywhere. They have a glass coating that protects them from small bumps, which makes them quite resistant. They are usually mounted on an aluminium profile that allows the creation of a ventilation channel at the bottom of the solar panel. This provides better ventilation in the solar panel (something essential). When installed on the roof of the van, it protrudes about seven or eight centimetres, slightly worsening the aerodynamics of the van. Its weight is higher than the weight of flexible solar panels. The advantage is that they are cheaper than flexible ones and that their performance is usually superior.

Monocrystalline or polycrystalline

Both panels are made of the same material (silicon), the only difference is their manufacturing process. In real terms, there is no major difference between one type of crystal and the other.

The monocrystalline crystals are better for cloudy days as they are better able to absorb radiation in the environment. The polycrystalline crystals on the other hand are better when the sun shines directly, as well as having improved performance at high temperatures. The polycrystalline plates are slightly more expensive. However, as we say, the difference between one type and another is barely noticeable, so choose the solar panel with the best value for money you find.

Why you should use mobile solar power

One of the most widely used energy sources is the one emitted by a generator. But this one has several disadvantages, such as noise and exhaust gases that can affect the camp environment. Thanks to the creation of mobile solar power systems, many of these problems are no longer present. They are the solution to enjoy on your next outdoor adventure.

Benefits of using mobile solar power systems for vans

Noise Elimination

One of the main attractions of traveling by caravan is being in contact with nature and away from the noise, so the use of a combustion generator can disturb the tranquility of the environment.

Reduction of Emissions

Most campers are environmentally conscious. They know how to appreciate the environmental benefits of using renewable energy, are more respectful of the environment, and the health and well-being of living beings.

Cost reduction

The sun is a free and inexhaustible source of energy. Costs in fossil fuels will be reduced and the investment will pay off in a very short period.

Local regulations

As a general rule, the use of the generator is allowed in compliance with a series of usage rules both in hours and in decibel levels. Therefore, in terms of noise, photovoltaic is the solution to comply with said regulations.

Autonomy

Mobile solar power systems offer a greater capacity for movement. Since there is no limitation of having to find a connection point to supply the consumption of the motorhome, and consequently a greater feeling of freedom.

CHAPTER TWO

Introduction to mobile solar power

Mobile solar power is an ideal renewable source for obtaining electricity from motorhomes. Using the sun to produce clean energy economically - thanks to photovoltaic panels, is a practice that is becoming popular due to its practical, economic and environmental advantages. In the practical aspect, the electricity provided by the solar panels allows its interior to come alive: light, hot water, air conditioning, heating, televisions, radios, and any appliance that we want to use on the road when parking or camping.

The conversion process is very simple: the panels convert solar radiation into electricity thanks to photovoltaic cells, just like any other solar energy installation based on this same technology. In this case, it is self - consumption and it is profitable, a key aspect, especially when establishing a comparison between the amortization periods and productivity with the useful life of the same caravan, which can be in about 15 years, and expiration similar to that of the panels.

A gasoline engine or solar energy?

Until now, motorhomes used generators with gasoline or diesel engines, pollutants by definition and more expensive in the medium term, since solar panels require a small investment that has to be amortized. Its placement through kits tailored to the needs allows electric autonomy, although its installation may require the help of professionals.

Logically, the number of solar panels will depend on the energy we need. A medium-sized solar panel can provide just over 200 watts of power, so we will have to add the necessary ones to meet our requirements. Also, we will take into account the energy that batteries can store for the night or gray days, and we can also help ourselves with a solar thermal system to heat the sanitary water.

Hot water with solar thermal energy

Indeed, in addition to photovoltaic self-consumption, solar thermal energy is a great ally for the generation of renewable energy in this type of vehicle. Fortunately, the installation of a solar thermal system is easy to apply in a caravan.

Specifically, flat solar collectors or thermal solar panels take advantage of solar energy to generate heat, applying an identical system to that used to provide hot water in homes. Its appearance is similar to that of photovoltaic panels and the volume of water that allows heating will depend on its total absorption area. In the case of motorhomes, a sensor adapted to that of the same motorhome will be required, seeking performance enough to meet the needs as tightly as possible.

Motorhome with photovoltaic panels

Its use could even extend to underfloor heating, in addition to serving as a shower and washing dishes. Beyond the number of advantages that we can achieve, energy independence changes the concept of traveling. It allows, for example, excursions to places further away from the madding crowd, far from civilization, without giving up a minimum of comfort and, very importantly, also taking care of the environment.

In addition to photovoltaic and thermal solar energy, the nomadic life practiced by the users of these houses with wheels can take advantage of their moments in the open air to get even more out of the king star. The solar cookers are another interesting option, both to preserve the productivity of solar panels for other uses and even to store energy in batteries, which will then be very useful on cloudy days.

Advantages of using mobile solar power

Mobile Solar power is renewable. You will never have to worry about running out of electricity for the use of electrical appliances. The sun is a constant power source, which means it will always be there every day.

Solar energy is part of the environment. Compared to fossil fuels that emit greenhouse gases, carcinogens, and carbon dioxide, solar cells do not emit anything into the air.

Mobile solar panels are very reliable. There are no delicate parts so you don't have to worry about replacing any part. Most people generate electricity for 1000 hours with little or no maintenance.

Solar cells make no noise while generating power. It is the only renewable energy source that is completely silent.

In the long run, solar electricity is cheaper than buying power from the power company. There is a start-up cost, but then you start paying for yourself. Once the break-even point is reached, all that remains is profit. Compare this with paying a monthly bill and not getting a return on investment.

Wide varieties of solar panel systems are available. Some can cost tens of thousands of dollars, and some cost only a couple of hundred. This means that anyone can purchase solar energy.

No need to connect to the mains. You can be completely self-sufficient and live off the grid. Imagine not paying another light bill.

Excess electricity is sold. If a solar power system is built enough to supply itself and there is still a surplus, you can turn your electricity meter upside down.

Solar technology is constantly improving. Solar installations are increasing by an incredible 50% each year, most of which are mobile home systems. Making solar energy more in demand and therefore cheaper on a larger scale, higher quality, and advances in its efficiency.

Explaining The Components of Mobile solar power system

Photovoltaic panels

The plates are the most visible part of solar installations. It is the component that you will put on your roof, in your parking lot, or on the ground to be able to capture the sun's radiation and convert it into electricity. Each board has a characteristic voltage and voltage range according to which the boards will have to be connected. Electricity produces it in direct current.

There are different technologies to make photovoltaic plates. The most common is crystalline silicon, either mono or polycrystalline.

Monocrystalline silicon plates are those that make wafers from a single silicon crystal. These have a somewhat higher performance than other conventional technologies, yes; they have more losses at high temperatures. The yield of the cells in the laboratory can exceed 20% in some cases, but assembled in the modules are around 17.2% due to the polygonal cut that the wafers must have.

Polycrystalline silicon plates are similar to the previous ones, only the crystallization process is carried out somewhat faster and several crystals are formed. This will give places to plates in which different shades of blue can be seen. The performance in the laboratory is lower than the monocrystalline ones, around 18%, but in the panels, since the wafers can be cut square, the performance difference is reduced, since they are around 17%. The performance of these panels in times of high temperature is better than that of polycrystalline panels and the price is usually somewhat lower.

To improve the performance of crystalline panels, reflective crystals have been used in some solar parks to increase incident radiation and create concentration panels. This solution is not widely used because excess radiation has been shown to end up damaging photovoltaic cells prematurely.

There is a third type of silicon panel: amorphous or thin film. These panels crystallize quickly without allowing crystals to form, which gives the module a continuous appearance. They have a much lower performance, not exceeding 12% and better behavior at high temperatures.

When architecturally integrating panels in vans, amorphous panels are often thought of because of the uniformity of color and because they can be made in flexible or adapted forms. There are amorphous panels that are not silicon, they are made of other metals such as Cadmium Telluride or Indium and Gallium Copper Selenides (CIGS) which must have a special treatment at the end of their useful life for using components that can become toxic if they are not managed well.

Crystalline panels can also be made translucent between layers of glass, slightly separating the photovoltaic cells.

Charge regulators

It is used in autonomous or isolated installations where batteries have to be charged directly. It is the equipment that is in charge of controlling the state of charge of the batteries and regulating the intensity of the charge to lengthen the useful life of the batteries. This equipment controls the current input coming from the solar panel to prevent overcharges and also to prevent the batteries from discharging more than they should.

Usually, two classes of charge regulators are used, the PWM, and the MPPT or maximizer. Depending on the type of plate we use, one or the other should be used. The difference is that the PWM works with a fixed voltage and plates that provide this voltage must be installed and, instead, the MPPT works looking for the maximum power point to maximize the intensity provided by the plates.

The panels used with the PWM are small panels, called 12 or 24 V, 36, or 72 cells. With the MPPT any plate can be used, so the panels used for solar parks or 60-cell grid installations are usually used since they are cheaper as they are more common. In the market it is already easy to find inverters that integrate the charge regulator in the same device, either PWM or MPPT, thus facilitating installation.

Inverters

At this point, we have to divide the section into two types, the battery inverters, and the grid inverters. This is because they have different use and operation even though they are called the same and both convert direct current to alternating current.

Electricity is normally consumed in the form of alternating current and solar panels. As we have said, to produce it in continuous direct currents, therefore, it must be transformed. A battery inverter is one that takes the current from a battery, at a certain voltage, and passes it to alternating current, either in single-phase or three-phase depending on the size, usually, single-phase. That is, it creates the right wave.

Some inverters can incorporate the charge regulator, as explained before, and also act as inverter-chargers. The inverter chargers also serve the opposite case, converting alternating current from an auxiliary source into direct current to store in the batteries.

This alternating current can come from a generator set, from the electrical grid or a grid inverter. This equipment will be used whenever there are batteries, either a network connection installation or an isolated installation.

The inverter networks are those that connect directly to the panels and generate an alternating current waveform identical to that of the network they are connected to. These inverters cannot function if there is no other network to which it is connected, which may be the public network or another one generated by a battery inverter or a generator. These devices are limited according to the country where they are connected in frequency and voltage so that they always make a wave the same as the one found.

Grid inverters are used in solar parks for the sale of electricity, in self-consumption installations, and isolated installations with an AC-SIDE connection. There are now some hybrid inverters on the market that unify all systems and make installation easier.

Batteries

Batteries are the elements where the energy produced by solar panels is stored during daylight hours so that it can be used when necessary, at any time. The batteries always store at a certain voltage, 12, 24, 40 V ... and have a maximum capacity that is usually measured in Ah.

Conventional batteries until now have been Lead Acid in liquid or gel form, which can present stability, maintenance, or durability problems. Today, lithium batteries are prevailing, also helped by the development of the electric car and other materials are coming to the market with very good benefits.

Purchasing Your Mobile Solar panel

Buying photovoltaic solar panels for your caravan, motorhome or camper is the best way to charge your batteries and gain autonomy by generating your electricity when you are away from home. Once installed, they provide clean, quiet power that you can enjoy anywhere.

But before buying panels, batteries, and components, you should consider the best type of solar panel that suits you, since there is a great variety. Some solar panel configurations are more efficient than others, and each installation will have different needs to meet. Before purchasing, you need to understand the different types of solar panels for caravans or camper vans.

There is a wide variety of rigid or flexible photovoltaic solar panels or panels; monocrystalline or polycrystalline; etc. Knowing the details of each type of panel will help you in your purchasing process and know which one best suits your needs.

Rigid solar panels

Rigid plates are solar cells mounted under tempered glass. They range in size from huge residential panels to small 20-30 watt device chargers. Most of the plates are mounted on an aluminum frame and designed for long-term outdoor use. They are designed to withstand hail, sand, and wind. The glass is scratch-resistant, which is best for long-term light efficiency. They are also easy to clean and in cold weather, they can withstand an ice-scraper.

In general, the price per watt of rigid plates is cheaper and they have a good guarantee (more than 10 years). Also, thanks to its durable frame, they are easier to install and direct towards the sun.

Pros:

- More resistant
- Cheaper than flexible
- Easy to face the sun
- The frame allows for better ventilation
- Better performance

Cons:

- High weight
- They can protrude from roofs and affect aerodynamics
- You will need a structure to fix it to the ceiling
- You will have to pierce the roof several times to secure it

That said, rigid plates are one of the best options since they have better performance and durability. The best rigid solar panels for caravans or campers

There is a wide market for rigid solar panels from various brands. In general, and as the plates are relatively simple devices, there is not a great difference between the quality of the materials used in their construction. But if we have to choose a brand, our recommendation would be Renogy, ECO-Worthy, or Enjoysolar. The three brands are very similar in terms of weather resistance, the efficiency of solar cells, and the warranty.

Many of these solar panels are intended for use in recreational vehicles and boats. They are lightweight and have drilled holes for mounting. These plates have a sturdy aluminium base that will allow you to use them as portable or ceiling-mounted plates. One of the benefits of the aluminium structure is that it keeps the solar cells elevated, which favors the airflow between the plate and the roof. This will keep the plates cooler and increase their efficiency and life.

All the components that come with these boards, from the junction box to the MC4 connectors, are waterproof. This means that you can travel under adverse weather conditions without fear of damaging the system.

Flexible solar panels

Flexible solar panels are made of flat molded cells with a protective plastic layer on top. Because they don't have a frame, they're low-profile and can bend to shallow curves, like the roof of a van. These panels are lighter than rigid plates. In contrast, the softer plastic, which is used in the manufacture of these panels, is more prone to scratches on its surface. However, due to this flexible nature, they are less likely to crack due to a large impact.

In the beginning, bending the panels too much tended to cause problems with internal connections and even led to a short circuit between the cells. However, these problems have been improving in manufacturing processes. For this reason, the warranty on flexible solar panels tends to be significantly shorter than rigid panels.

Flexible panels are generally more limited to sun exposure because they are not independent. Their positioning is determined by the surface to which they are attached.

Pros:

- Low profile
- Flexible up to about 30º
- Light
- Easy to assemble

Cons:

- Prone to scratches
- A shorter life
- More expensive the watt price than rigid plates

Solar panels work best when the entire panel is receiving constant light. With flexible panels, the more the panel bends, the less efficient it will be because part of the solar panel will receive less direct light.

Flexible solar panels are best if:

- You don't want them to be seen on the ceiling
- When the only available surface area is significantly curved
- If you don't want to make a big construction or drill holes in the roof

As with rigid solar panels, there is also a wide variety of flexible solar panel brands on the market. Many of these brands share features like warranty, efficiency, and extreme weather resistance.

Portable Solar Panels / Solar Chargers

Portable solar panels, solar chargers, or solar cases are separate, easy to fold/unfold units when you need electricity. These devices are a great option if you're looking to get a small amount of power, like charging phones or a laptop on your travels.

Portable solar panels

Pros:

- Easier to install
- They do not require a ceiling installation
- Can be positioned for the best lighting
- Adaptable to any type of vehicle
- You can place them in the sun while your vehicle is parked in the shade.

Cons:

- Prone to scratches
- A shorter life
- More expensive the watt price than rigid plates

Types of solar cells, monocrystalline vs polycrystalline

Photovoltaic solar panels are made of smaller cells which in turn are made of crystalline silicon. Depending on the manufacturing process, two different types of silicon crystals are obtained: monocrystalline and polycrystalline.

Monocrystalline-polycrystalline cells

Monocrystalline plate Polycrystalline plate

Although there is talk that one type of glass is more efficient than another, the truth is that 100W is 100W in both monocrystalline and polycrystalline cells. Where the efficiency of each type of cell is relevant is with size.

The basic differences are:

Monocrystalline plates are more efficient, take up less space, and tend to be slightly more expensive.

Polycrystalline plates are cheaper and larger, per watt, than monocrystalline plates.

Both types of cells share the same attributes of longevity and durability. You really can't go wrong with either option as long as you take action and make sure it meets your energy requirements.

Requirements for installing a mobile solar power system

If you are considering starting a solar installation in your caravan or camper van, a good option is solar kits. Solar kits are an easy way to get all the necessary components (plates, regulator, connections), so you can start generating your electricity with solar panels.

The main differences that you will see in the solar kits are:

- Solar panel type (monocrystalline or polycrystalline)
- Solar output power (in watts, W)
- Charge Controller Type (PWM or MPPT)
- Batteries (included or not)
- Power inverter (included or not)

The most efficient kits come with MPPT charge controllers and monocrystalline panels. The cheapest and tightest solar kits generally have polycrystalline panels and PWM charge controllers.

The main difference in the type of solar panel, for practical purposes, is the surface. Monocrystalline panels, being more efficient, are slightly smaller than polycrystalline ones. But with the advancement of solar technology, the efficiency between these two types of panels is negligible. Both work effectively. Remember: 100W is 100W no matter what type of panel you buy.

MPPT charge controllers are more efficient than PWM. However, you will only benefit from them if you use solar installation daily. For sporadic uses, a PWM controller will serve, in addition to being cheaper.

Chapter Three

Selecting a solar panel

Sizing You Solar Panel and Battery

In the case of mobile photovoltaic solar energy installation for a van, sizing is key. In this way, we can self-supply in a more efficient way at an energy level and adjust much more economically. The power of a solar panel is about the sum of the daily consumption by the elements of your electrical installation. This consumption must be found in watts/hour (Wh) and for this, a very simple calculation is made.

We simply need to know the power of each element in watts (W) and multiply it by the estimated daily operating hours of this element:

[W (appliance Watts) x h (hours / day) = Wh / 24]

As an example, we are going to put the installation of an average van to give you an idea of what you should take into account.

Calculation of the daily consumption of an electrical installation of a van

The first thing you have to do is find out the overall consumption of the electrical installation of your van. You must take into account all the elements that you are going to install. Also if you are going to put any item of sporadic consumption such as USB charging points, etc., we advise you to make an estimated calculation of your daily consumption.

As we have said, we leave you the example of an average van to serve as an orientation and how we consider what consumption we produced.

With that said, we use the watt-hour per day formula we've seen previously:

[W (appliance Watts) x h (hours / day) = Wh / 24]

To determine this value separate the summer and winter consumptions by the consumption of the heating of your van.

Summer Consumption

Fridge 550Wh / 24h *

Water pump 54W x 0.5h = 27Wh / 24h

LEDs 5W x 4h = 20Wh (x 4pc) 80Wh / 24h

Estimated consumption usb plug 5W x 3h = 15Wh (x 2ud) = 30Wh / 24h

Total summer = 680Wh / 24h

* Data extracted from the instructions of an average OS fridge. Refrigerators only consume a third of the power while connected (about 8 hours a day).

Winter Consumption

Total summer = 680Wh / 24h

Heating 5000W x 2h = 10000Wh

Total winter = 10680Wh / 24h

Sizing of the solar panel in a van

First, you have to do the calculation between what the panel generates and what the installation of your van consumes. As this last data we have already extracted in the previous section (680Kwh / 24h), to calculate the energy demand that you can generate with your photovoltaic panels the following formula applies:

[Average daily irradiation of the area where you live x maximum power of the installed solar panel]

Multiplying these values you get the energy that your solar panel can provide you throughout a day.

In this case, we use the average daily irradiance value:

4.7 kWh (BCN value) x 150W (maximum installed solar panel power) = 705kWh / 24h

Knowing the result of the consumptions generated by all the elements of our van (680Kwh / 24h), we conclude that, in summer, our solar panel for the van is sufficient.

As you can see, in winter the electrical installation of our can not be self-sufficient only with the solar panel for the van since the heating triggers consumption (10680Wh / 24h). For this, more solar modules and more power would be needed to supply winter consumption. In other words, the greater the demand for consumption, the greater the number of panels we must install. In our case, if we are several days in the same place, we would need another charging source such as the van's alternator to charge the secondary battery.

How to size the capacity that your battery must have

Once you know the daily consumption of your solar installation, the following question comes: What battery do I put? In this section, we will explain the type of battery you need according to your consumption.

The data that we will need for our calculation is:

1. The total daily consumption that we have previously calculated (680Kwh / 24h)

2. The days of autonomy. This factor will depend on the use of the installation. In case of daily use we will use 5 or more days and in case of sporadic use, weekend or holidays, we will use 2-3 days.

3. The depth of discharge. In the case of gel or AGM batteries, which are the ones used as secondary batteries, a 70% (0.7) depth of discharge is considered. In the case of open lead, it is considered 50% (0.5).

4. The battery voltage, which in our case is 12V.

5. Losses due to temperature and installation work. Each work and performance of both mechanical and electrical equipment implies a loss of energy due to temperature, friction, etc. So we have to take into account a percentage and we are going to give it a 15% loss.

Taking these data into account we obtain the formula for the capacity in battery amps:

$$Ah = \frac{Wh/24h \times Días}{Pd \times V} \times \%\ Pérdida$$

Transferring our values we obtain that:

$$Ah = \frac{680 \times 2}{0{,}7 \times 12} \times 1{,}15 = 186{,}20 Ah$$

This indicates that the battery should have a minimum capacity of 186.20Ah, so a 200Ah battery would be sufficient for our installation.

How to Install a Mobile Solar Power

Installing a mobile solar panel on a van is pretty straightforward. At first, it might seem like it is going to be more difficult for us than it is. Below we explain the steps to follow in your installation.

Tools to use for the installation of a solar panel in a van

These are the tools that you need for the installation.

- Electric drill
- Battery drill
- Cable cutters
- Tester
- Shear Cable Crimper
- Plier Terminal Crimper
- Tin soldering iron
- Metal drill bits
- Drill bit set

- Squad
- level
- Measuring tape
- Carpenter's pencil

Function test of the solar panel plate

First of all, you have to verify that your 12V solar panel works perfectly and that it gives you the correct voltage. For this step, you are going to need a tester that will allow you to know the working voltage of your solar panel. This is usually between 15 and 19V.

In daylight, connect positive and negative testers with positive and negative solar panels, and if it gives you a reasonable voltage for your solar panel, you can install it.

Location of each item

Previously, it is essential to determine the location of each element in its corresponding and useful place. This will help you to know the amount of cable you need, if the placement of each component bothers something, etc.

Pass the wiring

Once we have a clear location and know the amount of cable, we can wire. We will have to pass the wiring of our solar panel kit from the solar panel to the regulator, from the regulator to the battery and from the battery to the inverter.

In the section of cable from the plate to the regulator, we must make a hole in the sheet metal of the van to pass the wiring. This hole must be large enough for two 4mm² section cables to pass through. Then it is recommended to seal with silicone and cover with a grommet.

In the positive cable between the solar panel and the regulator, we install a fuse. This is very important to protect your installation against any circuit overload and so that there are no electromagnetic interferences. Sometimes, the hybrid inverter already has the function of a charge regulator, so you can directly connect the board to the inverter. In the same way, install a fuse (15A) in the positive cable. Finally, simply connect each element to different terminals. The use of various tools is essential for this step.

You will need a shear cable crimper for the placement of the battery connection terminal (section 25mm²) and a plier terminal crimper for the smaller cable sections (4mm²). On the other hand, you will also need a tin solder for MC4 connections. The MC4 connectors are electrical connectors used for connecting solar panels.

How to connect solar panels in series or parallel

The need to connect our solar panels in series arises because we want to obtain higher voltages than those currently offered by the panels. Thus, for example, to obtain 48v we will need to connect several panels in series because there are no solar panels that offer that voltage. Or because the inverter's working point asks us for a minimum working voltage or voltage

Depending on our objective, solar panels can be connected in series if we want to increase the system voltage, or in parallel, once the working voltage is reached if we want to increase the amperage to have more power. The way of making the connection may affect the design of the photovoltaic installation.

Connect solar panels in series

In this case, the connection between the panels is much easier. We will only have to connect the positive terminal of the first panel with the negative terminal of the following panel. With the connection of solar panels in series, the working voltage (Volts) is multiplied keeping the current intact, just opposite to the parallel connection. In the series connection, you must be very careful not to exceed the maximum voltage for which the charge regulator or inverter is designed, either from the grid or isolated to which we connect these solar panels.

This method is typically used for large/powerful solar panels. The most common is that they are arranged in a series of 2 or 3 panels because they have to be connected to a regulator of the MPPT type. In the case of inverters with isolated systems, grid-connected inverters usually require more panels in series, such as general rule more than 10 out of 72 cells.

Important: To correctly connect the panels to the regulator or the inverter, we will look at the panel's technical data sheet, and we will observe the Voc (Open Circuit Voltage) value, multiplying this value by the number of panels shown and thus comparing it with the Max PV value. VOC (the maximum open-circuit voltage supported by the regulator/inverter). With the data obtained, we will clear our doubts and we will be able to choose how many panels in series we can connect.

Connect solar panels in parallel

If we want to add more power to the series, but we cannot add more panels in series, we will make another group of panels the same and connect them in parallel, we join positive and negative terminals of the series in a single connector

This type of installation is carried out by joining all the negative poles of the plate/series and on the other hand by joining the positive poles. In this way, the parallel connection of solar panels implies a sum of charging current (Amps), however, we maintain the same working voltage (Volts or Voltage). For this type of connection, we can use a wiring hub. The power distributor will allow a simpler and safer connection. For associations of more than two (2) panels in parallel, it is necessary to use fuses to protect the series.

Panels must be identical or as similar as possible for best performance. As in the series of modules, we must verify that the max PV (W) value of the regulator or inverter is not exceeded, nor the maximum current that the inverter/regulator can handle.

Connect panels in series and parallel (mixed)

This type of combined connection is used to increase both the current and the voltage of the system. The most important thing is not to exceed the maximum voltage allowed by the load control or the maximum power. It is normally used in installations where 4 or more 60 or 72 cell solar panels are connected. In this case, we seek to multiply the voltage and total amperage of the installation. MPPT charger inverters are used for this type of installation. All series in parallel must be the same in the number of panels.

Is it better to connect solar panels in series or parallel?

Depending on the magnitude of the photovoltaic installation, the working voltage of the batteries (in case of isolated self - consumption), the output voltage to the inverter (in the case of large installations), and the specific specifications of the solar panels. They will use one or the other, or the combination of both.

It is very important to make the right decision to obtain adequate performance for your type of installation and obtain the desired voltage.

However, whenever the system allows it to work in series, it always reduces losses. If it is technically feasible to connect, for example, two modules in series or two in parallel, you should always opt for two in series.

What differences exist between the types of connectors (MC4 and SAE)

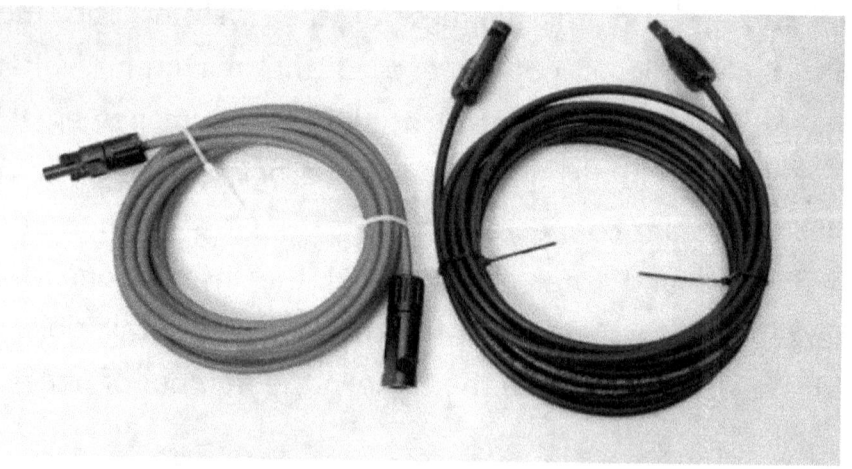

Multi-contact MC4 connectors are used to guarantee a correct connection of the solar panels. There are single, double, and triple connectors, the latter is used to put the solar panels in parallel.

The difference in conductivity between a solar panel connected by the MC4 connectors and a solar panel connected with SAE connectors can be very important, especially taking into account the passage of time and the effect of the sun, wind, and water on the insulating tape or other connection mechanisms. The connectors for solar panels guarantee correct sealing of the connections and avoid future problems in solar installations.

There are several types of connectors: Tyco, MC3, MC4, etc. The most common for their ease of assembly is the MC4 and SAE. These are the best connectors that your solar kit can carry.

These connectors will make the connection and security of the panels in your photovoltaic self-consumption even easier. The MC4 and SAE connectors are installed with the help of a crimping tool for 4 and 6 mm cables, which can be obtained at any hardware store. If required they can be sent crimp (connected to the cable).

Best practice for a water-tight cable installation

Step 1: Liquid Electrical Tape

Make sure the product has been mixed and use the brush provided to apply it to the exposed joint. A minimum of two coats is recommended and wait 10 minutes between coats. Let dry for at least 4 hours. This product can be applied to solder joints, butt connectors, male and female connectors, whether they are connected or whether they fit into a fixed connection, screwed into connections, etc.

Step 2: Self-fixing silicone tape

Try to keep the opposite end of the cable fixed, attached to something or having an assistant who facilitates the application procedure. Use a razor to cut a length of ribbon you will need and apply a 2/3 overlap when wrapping the ribbon around the exposed conductor. The band will stretch up to 3 times its length, the more the stretch will cause greater adhesion. It takes about 24 hours to melt completely, this tape can be applied to various connections, such as solder joints, butt connectors, male and female spade connectors, which are connected, etc.

Step 3: Heat-shrinkable tube lined with adhesive

Choose a heat-shrinkable tube with a diameter suitable for the exposed conductor. Cut an appropriate length; make sure it overlaps the existing insulation. Use a heat gun to heat the thermal shrinkage, which will shrink around the connection and soften the adhesive. Once satisfied, allow it to cool too much. Unfortunately, this product is only limited to soldered connections between two ends of butt connectors or cables.

Chapter Five

Introduction into boats mobile power system

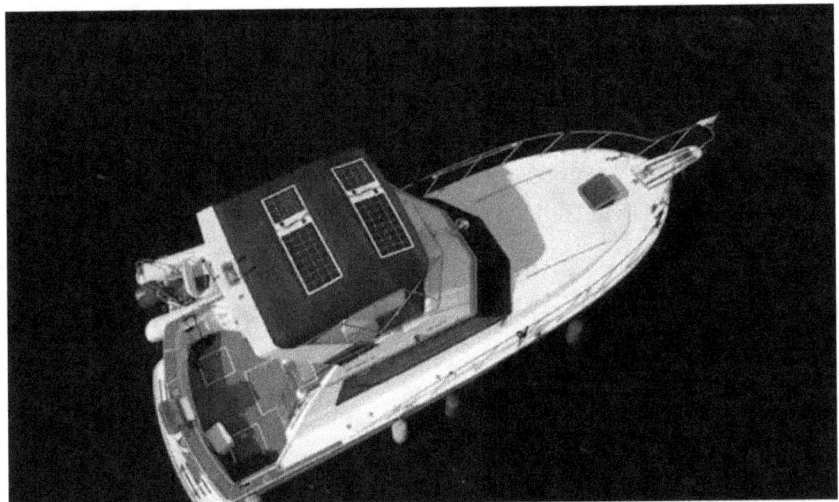

Sailors know the importance of resource management. One of the most important in terms of navigation is electrical energy. Whether for a sailboat or a motorboat, electricity is essential for the navigator's equipment. The solar panels for boats participate significantly in energy saving, and they are fast becoming a "must" on any vessel.

There are many parameters to consider when choosing the best installation solution. Consider the following characteristics:

- the size of the boat
- the level of equipment
- the design
- the usual navigation conditions

It is from these characteristics that we will choose the location, type, and power of the photovoltaic solar panels to be mounted.

The solar kits for boats are mainly composed of flexible photovoltaic solar panels. They are usually mounted on the deck or the gantry of the boat. Their flexibility makes them easy to integrate into the design of the ship. The thin panels, however, resist trampling. They can be removable or permanently fixed. Like solar panels for motorhomes, there are standard ready-to-use kits.

In general, the kits are composed:

- a solar panel
- a charge regulator
- UV and weather-resistant cables
- roof hatch
- glue tubes

It can appeal to professionals of solar panels for boats for a custom installation. A preliminary study is carried out before the design and manufacture of the system. It integrates the energy requirements of the boat (size of the panel), the aesthetic constraints (integration about design, size), and the various specifications of the owner.

Advantages of using mobile solar power for boats

In the current nautical field, it is important to have a DC power supply (direct current) to replace the energy of the boat's batteries. Electricity is necessary to use the navigation systems and radio on board, as well as for the lighting and use of household appliances and other electronic devices.

For long voyages, sailing ships, or ships where electricity consumption is high, it will be a good option to install photovoltaic solar panels that provide the batteries with extra daily energy to cover their electricity needs. In this way, it is totally or partially avoided that the propulsion motor must be used for a couple of hours a day to charge the boat's batteries, the energy of which will have been consumed. Therefore, photovoltaic panels are an excellent and major option for use on sailboats and large sailboats.

To install the solar panels permanently you can take advantage of the deck of the ship, or the bimini on small boats. If the surface of the roof is flat, it is advisable to install the classic solar panels made with safety glass and aluminium frame.

Quite a few years ago, solar panels had performance issues when there was shade somewhere on the panel or one of its silicon cells. Today this problem has disappeared since each panel has bypass diodes, which allow the solar panel to generate energy even if some part of the panel does not receive solar radiation.

This fact is very important in solar installations on boats, since there may be a lot of shadows from the boat itself in motion. Current diode manufacturing solves this problem and allows the marine solar installation to get the most out of the sun. It also allows it to take advantage of any available space within the limited surface area of the boat.

The number of solar panels to be installed will depend on the function that we want to give to solar energy. It can be a battery charge maintenance facility, widely used in small or medium boats to maintain the level of charge during the long seasons when they are unattended. This avoids finding a dead battery and having to buy a new one. Also, It can be an installation for the daily basic operation of the essential devices for safe navigation. This includes route and communication devices, in which the 500W solar boat kit can be used.

Finally, it can be used to cover large installations, in addition to navigation equipment, lighting consumption, and household appliances, such as the refrigerator, or for leisure such as television. Once the plates are installed, their energy will pass through the charge regulator. The charge regulator has the functions of gradually and optimally storing the energy generated by the sun in the batteries. It will also have a very useful function since it will allow us to know at any time the percentage of charge level in the accumulators of the boat.

Thus, all the energy that can be obtained from the sun will avoid the use of the engine and will allow significant fuel savings. Other advantages will be the end of the noise of the generator and the energy independence of the ship.

How to install mobile solar power for boats

To install a mobile solar power system, you need one or more solar panels, a charge regulator, and one or more utility batteries. But the first step before choosing the material is to make an approximate calculation of the daily consumption that we will have (always being generous).

In this case, we will use a small sailboat without too much electronics and consumption of about 50 amps a day.

Types of solar panels

Once the calculation has been made, you must choose a suitable solar panel. There are two types on the market, Monocrystalline and polycrystalline. The former is more efficient (smaller size for the same power as a polycrystalline), and the latter is cheaper. Some come in a rigid or flexible format, which is much thinner and lighter and can be installed on deck.

You can opt for a 160w rigid monocrystalline panel with a 12v working voltage, capable of providing up to 9A per hour. When multiplied by approximately 8 hours of sunlight, it would theoretically cover a daily consumption of 72A. This may not be completely accurate since there are infinite factors that limit the charge of our batteries. These factors include the inclination of the panels if there are shadows on them or the simple fact that it is cloudy. Therefore, the plate must always be oversized. In our case by 30%. You also have to decide if you want to have photovoltaic cells inflexible or rigid panels.

Regulator

The charge regulator is the one that controls the amount of energy between the panel and the batteries. You can choose a PWM regulator, which is mostly used for panels with voltages between 12 and 24 volts. They are the simplest and cheapest regulators. The other type of regulator is the so-called MPPT, they are usually used in higher power panels that work with voltages of 30 volts.

Batteries

Batteries are measured in amp hours of consumption. These must be able to satisfy the energy demand you have when there is no sunlight. Also bear in mind that they should not be downloaded more than 50% since it would greatly shorten their useful life.

Basically as mentioned before, there are three types of batteries, the traditional Lead Acid, Gel, and AGM. The main difference is that those of Gel and AGM accept slower and deeper discharges than those of Lead Acid, which is what you need in a service battery. The difference between Gel and AGMs is that Gels accept even slower discharges and have more durability. However, the downside is that it needs a much more precise charging voltage to prevent damage and is priced higher. You can choose a single 180 ah AGM battery, more than enough size for consumption.

Installation

The electrical installation is really simple, once the appropriate cable diameter has been chosen based on the cable length and the maximum intensity provided by the panel, the panel is connected to the regulator, and the regulator to the battery. The main problem is where to mount the solar panel. It can be installed practically anywhere and on the market, there are all kinds of supports, but each boat is a world.

You can devise a light yet strong stern bracket, which could also be oriented slightly depending on the tilt of the sun. These are two stainless steel tubes held firmly on deck. It is placed on the aft balcony and reinforced by winds of steel cable. You can install a rubber tendon at the top of each tube and the plate above it. Thanks to the flexibility of the tendon, the plate can be easily oriented. To fix it in the desired position there is a rope at each end of the panel that is tied in a cleat mounted on each tube.

Chapter Six

Introduction to Small Homes Mobile Power System

Although there are currently more and more users who switch to new energies as a way to take advantage of the resources that the planet offers us, there are still many skeptics who wonder what are the true advantages of having these sources at home. Specifically, with solar energy, there is a lot of debate about whether it is the best option for the future to leave behind the other energy pathways that pollute more and damage more the environment in which we live. If you are in the group that still does not know if it is a good bet for your home, this chapter can help you solve the small (or big) doubts you may have about mobile solar power.

Advantages of Using Mobile solar Power System for Tiny Homes

- Once installed, you can start enjoying a regular and constant supply of electricity in your personal space and offices; without having to worry in any way about pollution, carbon footprints, or any other health risk.

- If you have your network design, then in effect, you are enjoying less dependency on fossil fuels and foreign oil.

- If you happen to reside/work in a region known for unpredictable weather conditions, then you may depend on clean, renewable energy, every calendar day; even cloudy days end up generating low energy levels.

- Unlike payments on your utility bills, a better return on investment can be achieved after installation.

- As these solar panels last for more than 30 years or more, without any daily maintenance or other costly supplies, they turn out to be a profitable option for you.

- People involved in installing solar panels as solar power installers as well as solar installers create more jobs for others in the industry, thus helping the economy profitably.

- The excess energy generated by the solar panels can be sold to the localized electricity company, as previously agreed.

- Once you start generating enough power for your home/office, you get the free-living grid capacity at all times, with consequent huge money savings on an annual basis.

- Panels can be installed anywhere from the top of your building to an open patio.

- The correct solar power generation system comes equipped with scalable batteries that store additional energy for use at night and other days in the future.

- Energy alone can be used for household power, water heating, and car runs as well.

- In general, solar energy has proven to be safe and more comfortable to use than its traditional counterparts (electric current).

With other innovations in the field, the systems involved for the purpose are proving to be more robust and accessible than ever, if you choose this system you will not be disappointed. Installing a mobile solar power system for your small home is one of the best investments. Helps you achieve maximum efficiency and reduces energy costs.

Components of Tiny Homes Mobile Solar Power System

A self-consumption photovoltaic installation for tiny homes contains very diverse components such as photovoltaic panels, optimizers, support structure, battery, solar inverter, bidirectional meter, monitoring system. In this section, we explain what each of these components consists of and what their function is in the installation.

Photovoltaic panels

Photovoltaic panels, also called modules or photovoltaic panels, are responsible for generating electrical energy from the light incident on them thanks to the so-called photovoltaic effect. The power of a solar panel is not the only factor to consider; its efficiency and long-term performance guarantee should also be considered. For this reason, a module with less power can offer better benefits than another with more power if these two parameters are higher since it can generate more energy on less surface area and its production will be less affected by the passing of the years.

Other external factors to the panel that will affect its energy generation are solar radiation (which in turn depends on the geographical situation), the orientation and inclination of the roof, and the possible shadows cast by elements such as chimneys or castles for example.

Charge Regulator/Power Optimizers

Power optimizers are components located between the set of photovoltaic modules and the inverter. It's objective is to improve the performance of the system by operating each module at its maximum PowerPoint.

When a panel of a branch or string of modules (set of panels connected in series) has a lower performance than the others, either due to factory defects, shadows, or other circumstances, the rest of the modules will operate at the same power as a said panel. This drastically reduces the overall performance of the installation.

One of the easiest ways to avoid this problem is by installing optimizers, to which each panel (or pair of panels in large installations) is connected, making each one operate independently of the rest. 1Another way to overcome this drawback is by using micro inverters, although it is a very expensive option but at the same time very flexible, allowing the number of panels in the installation to be expanded without limitations.

Both solutions also allow better monitoring of the panels, facilitating the detection of faults. They are normally attached to the back of the panel. It should be noted that its use is optional unless the investor requires its use.

Support structure

This component is in charge of fixing the panels to the ground or roof, there are two types:

Coplanar: they are used when you want to place the panel glued to the roof, it is the most attractive solution from the aesthetic point of view, they also allow you to take better advantage of the available space on the roof.

Triangular structure: they correct the inclination and orientation of the panel optimizing electrical production. Its use is essential in flat roofs, they can be fixed using screws or ballasts. Their price is slightly higher than that of coplanar structures and they take up more space because they make it necessary to leave space between rows of panels to avoid shading them.

Battery

In photovoltaic self-consumption installations, it is in charge of storing the surplus photovoltaic production for later use. It should be noted that it makes the installation more expensive and delays the recovery period of the installation to about 10 years. However, these systems provide certain advantages such as allowing the continuity of the electricity supply in the event of a power cut and better use of photovoltaic production.

Some brands integrate inverters and batteries in a single pack. They also include other additional functions such as the ability to predict the production and consumption of the next day, charging at night from the electricity grid when energy is cheap.

Solar inverter

It is in charge of adapting the current of the panels or the battery, according to the configuration of the system, to alternating current to be compatible with the electrical equipment of the home or business. It also stores information on photovoltaic production, network consumption, and the general state of the installation, making it possible to monitor all this information.

Bidirectional counter

It is the meter of the house, at present almost all of them are bidirectional. This means they allow both to measure the consumption of the house as well as what is dumped into the grid by photovoltaics.

Monitoring system

It is the application or web portal in which the inverter is registered. It is where it inverts its data every few minutes, allowing the user to monitor photovoltaic production, household consumption, and the status of the electricity through their PC, mobile, or tablet.

Protection equipment

They are necessary to protect the equipment and the home user against various incidents. The most common being fuses (short circuits), the thermomagnetic circuit breaker (overloads + short circuits), the differential switch (shunts), and the varistor (overvoltages).

All these devices are usually integrated into an electrical panel located next to the inverter itself. Grounding can also be covered in this section, which protects against unwanted contact between conductors and housings.

How to Install Mobile Solar Power for Tiny Homes

Before moving on to the procedure to install solar panels yourself, let's briefly recall how a photovoltaic installation works. Solar energy is based on the conversion of the light from the sun into electricity. Photovoltaic panels are designed with semiconductor materials and silicon cells which, in contact with the sun's rays, will transform solar energy into electrical energy. Once created, the energy can be transformed into electricity, either directly available through an inverter, or stored in a battery designed specifically for this purpose.

Positioning the Solar Panels

Overlay photovoltaic panels

The superimposition consists of placing the panels on a structure fixed to the roof utilizing rails. It is the most frequent installation in small homes. If this installation is possible for all types of covering, it nevertheless requires a slope of a minimum of 8 °

Integration photovoltaic panels

The integration is the replacement of part of the lining of your roof with solar panels. It is a very aesthetic solution that allows you to renovate your roof (insulation and waterproofing). The installation on the front of the photovoltaic installation is also possible if your roof is too small or poorly exposed. The panels are then tilted at 30 ° and fixed to your facade utilizing a lifting device. These can then act as a sun visor for your terrace.

The installation of solar panels on the ground is also possible. This requires that you have a sufficiently large plot. The installation is then fixed to a modular structure generally inclined at 35 ° and facing due south.

Finally, the panels on solar trackers are also placed on the ground and these adapt their orientation to best capture the rays of the sun.

Material required for a photovoltaic installation in

Sufficient solar panels to supply your home with electricity;

Of rails that will support your installation;

The cables required to connect the photovoltaic panels;

An inverter, which will be used to convert the direct current produced by the panels into an alternating current usable in your house;

An electric meter that will measure the amount of energy produced;

In some cases, a sealing system will be necessary;

Finally, all kinds of tools will be necessary (screwdriver, saw, ladder, protective equipment)

Mobile solar installation in small homes: the different steps to follow

Remove the roof covering at the location of the solar panels

If you want to integrate your solar panels into your roof, you must first remove its coating. Initially, it will be necessary to determine what will be the surface of said panels. Then you will remove the covering from your roof on a slightly higher surface.

Remember to take safety precautions when installing a fall arrest system and always be careful where you put your feet.

Installing the waterproof base

Also if you have decided on integration and once your coverage has been removed, you will need to install a screen under the roof or a waterproofing system. This will prevent water from infiltrating the roof but still allow air to circulate for perfect ventilation of the roof. This will avoid any risk of humidity in the house.

Fixing of photovoltaic panels on rails

This stage also concerns integration, superimposition, laying on the ground, and the facade. Indeed, this is the stage central and necessarily inevitable.

Concerning the installation of the panels on the roof (overlay and integration) the fixing of the rails to the rafters, using hooks is necessary. This is because they serve as supports on which the solar panels will be fixed. This step is therefore very important and must be carried out with meticulousness and attention.

On the ground, the solar panels are placed on rails fixed to a supporting structure firmly anchored in the ground. On the facade, the rails are fixed to a bearing structure fixed to the structure of the facade walls. On a flat roof, the solar panels are also placed on rails that are fixed to a supporting structure allowing their tilting.

Lay connections between siding and panels

Here again, this is a necessary step if you have decided to integrate the installation into your roof. After removing the coating, the waterproof base placed and solar panels, it will ask the flashings. The fixing of the arguments on battens is quite simple and is done using a screwdriver. The installation of these parts makes it possible to make the connection between the roofing materials of the roof and the solar panels to ensure a good seal.

To finalize your installation, you will need to connect the inverter to the installation and the meter. For this step, it is essential to know how to read an electrical diagram.

Indeed, some panels will have to be connected in a series, others in parallel. This process is important for the proper functioning and longevity of your photovoltaic installation.

When this step is completed, your solar panels will be operational and you will be able, once the connection to the grid manager has been made, to benefit from the green energy produced.

Chapter Six

A simple guide on how to maintain a mobile solar power system

Mobile Solar photovoltaic installations are constantly subject to external agents and temperature changes that affect both the panels and the electrical connections that comprise them. For all these reasons, it is important to have good facilities maintenance.

Maintenance depends on several factors such as:

- The complexity of the installation,
- The climatology of the area
- Even the environmental contamination existing in the installation location.

Pollution and dirt settle on top of the panels causing a decrease in the radiation filter causing the production of the plant to be reduced considerably. Although most of the dirt comes from dust, the particles created by man-made pollutants such as carbon by ions are smaller and cause considerable energy losses.

How is the maintenance of the solar panels carried out?

It is essential to ensure the correct operation of the panels and guarantee the longest useful life of the installation that has adequate maintenance of its components. To perform good preventive photovoltaic maintenance in solar installations, you must:

Perform maintenance of the photovoltaic field (solar modules):

It consists of cleaning any type of object, dirt, etc. that can affect the correct production of solar panels. As we have previously commented, accumulated dust or pollution residues must also be eliminated.

Cleaning should always be carried out with non-abrasive products, thus avoiding damage to the panel. You can use products such as osmotic water, soap with a neutral PH, etc. and in any case following the manufacturer's maintenance recommendations.

Visual inspection of photovoltaic panels

In addition to the cleaning of the photovoltaic field mentioned in the first instance, a visual inspection of the panels should also be carried out in search of anomalies.

Checking the structure that supports the panels

The support structure of the photovoltaic panels is usually made entirely of aluminium profiles and stainless steel hardware, so they do not usually need anti-corrosion maintenance. However, it should be verified that there are no deformations or cracks, the tightness of the cover and that the state of fixation of both the structure to the surface and that of the modules to the structure is optimal.

Review of electronic components

Lastly, for the correct preventive maintenance of the installation, all the electrical components of the installation must be checked: direct current, alternating current, inverters, monitoring system, etc.

Active surveillance and telematic control of the installation must be carried out, in addition to verifying the components on-site. Clean or replace filters or any part that may be susceptible to error, as well as verifying the aging of all the components to carry out the appropriate corrective actions in each situation.

Maintenance of the accumulation system

It is important that in isolated installations where accumulation systems are available (solar batteries), the upper part of the batteries and terminals, as well as the connection terminals, should be cleaned regularly. The electrolyte level of the accumulation systems should also be monitored. If necessary, you should fill the electrolyte up to the level recommended by the manufacturer, since this will ensure the useful life of the batteries and keep them fully operational.

Likewise, it will be advisable to check with the density meter the state of charge of the battery and its capacity by measuring the electrolyte density. Periodically, you will also have to perform equalization of the batteries. This will restore them to their storage capacity, increase efficiency, and extend their useful life. This is achieved using a voltage overload applied in a controlled manner on the batteries to be equalized.

Troubleshooting Your Mobile Solar Power System

One of the most important factors that we must take into account is that the installation is done correctly for its proper functioning. To achieve this good operation, there are some aspects such as the location, number of panels, how they are connected, the geographical position of the house, among others, which ensure that we have an installation with correct performance. But what does this mean? It means up to a 50% increase in productivity and 30% less maintenance or possible failures or breakdowns. In the following article, we explain the possible problems you may have with a mobile solar power installation.

Types of problems that can occur on a solar panel

Solar panels and their components transform solar energy into thermal energy and electricity for use in homes and commercial establishments. In the same way, like many other installations used for energy production, solar panels can also suffer damage. The existence of defective components is easy to discover because it is indicated by the decreased performance of the solar installation.

One of the advantages of solar panels is that by presenting breakdowns and being repaired, they help in solving the most common problems affecting the modules. As long as the experts in the field carry them out on time and the user quickly informs the specialists about the breakdowns.

Prior knowledge of the common problems of the solar panel can alleviate the concerns of additional investments, about the new photovoltaic components. Below, we will describe some problems that may cause solar panels to have reduced performance or, failing that, it is either late for a possible repair:

1. Micro fractures and hot spots

Microscopic fractures, hot spots, and cracks can appear on the surface of the glass cells of the solar panel, and subsequently become larger or noticeable over time, affecting the effectiveness of the solar cells. Lamination, panel frame, and solar system waterproofing can remain in good condition despite cracks. The most common causes of these micro-cracks and so-called hot spots are the following:

Changes in ambient temperature or other weather conditions.

Some failure in the production of photovoltaic modules.

Mismanagement during the shipping process.

Deciding to repair these fractures and the hot spots on the plates depend on how the panels have been configured in specific installations. If the installation we have in the house is an integrated system to the roof, it must be completely dismantled, even if it is only a couple of panels that have low performance. It is worth cleaning and determining the level of damage on all panels before repairing broken glass panels.

2. PID effect

Potential induced degradation (PID) is attributed to voltage fluctuations that occur between the voltage generated by the solar panel and the solar panel's grounding. This leads to a variable percentage of voltage discharged into the main power circuit. Specialists in the installation and repair of solar panels will be able to correct this problem, and thus avoid degradation or aging of solar panels and achieve stabilization of their performance. Accumulation of dirt can also lead to lower energy output from solar panels and should not be overlooked.

3. Internal module damage

In some cases, solar panels can cause internal damage to your system. This damage is most often attributed to defective production or a selection of low-quality components. This may result in a result known as "Snail traces" and refers to delamination, discoloration of connectors or solar cells, yellowing of leaves placed on the back, the gilding of EVA (ethyl vinyl acetate) films, staining of photovoltaic cells, unwanted inclusions in photovoltaic laminates, burning of the back and/or front of solar modules, etc.

In most cases, experts remove the glass from the solar panel so that it can be repaired or changed. However, the type and extent of damage are best assessed by solar panel repair companies, and then the feasibility of the action is determined.

4. Loose wiring

The cables are responsible for connecting individual photovoltaic cells to inverters and domestic solar batteries. The smallest damage or failure that occurs due to wiring, will cause failed connections in the system and problems will arise in power generation due to imperfect or loose cables within the connections.

When conducting a fault assessment, solar panel specialists often use meters and other cable performance verification tools to assess such problems and provide timely and useful solutions.

5. Internal corrosion, delamination

Internal corrosion (oxidation) occurs when moisture penetrates the panel. Panels must be air and watertight. To achieve this, the components of the panels (the glass layer, the solar cells, and the back sheet) are vacuum laminated. Still, it may happen that the lamination process was not successful or was too short. Which may cause delamination during the operation.

Conclusion

As we mentioned earlier, a mobile solar power system is the perfect ally for the energy efficiency of your camper van or vehicle. Photovoltaic solar energy as it is often called is a clean energy source when it comes to impact on the environment. It produces electricity of renewable origin from solar radiation using a semiconductor device, the photovoltaic cell.

Mobile solar installation in a well-dimensioned camper allows you to have more freedom on your route since your van becomes self-sufficient. For this reason and others that we are going to explain, solar panels are a key part of the electrical installation of a van. This is even important if you are going to lead a nomadic life. Parked or driving, cloud or sun, it doesn't matter, as long as there is daylight, solar panels continue to produce electricity.

Before you begin installing the mobile power system, remember to follow each step given in the guideline. Also, ensure you observe proper safety precautions to avoid accidents or injuries that may occur. So what are you waiting for, pick up your tools, get your mobile solar kit, and start installing your own mobile solar power system?

About the Author

Larry Barone has over 10 years in setting up renewable energies systems and technological research for clean and more efficient energy. He works so hard to simplify the installation and maintenance of complex solar power systems, making it very easy for hobbyist to install them at homes, vans, RVS, boats etc.

He lives in New York, USA. He is happily married with two kids.

www.ingramcontent.com/pod-product-compliance
Lightning Source LLC
Chambersburg PA
CBHW070253220526
45465CB00004B/1601